Medardus Ramsauer

Hydrocarbon Reservoir Simulation

AF135942

Medardus Ramsauer

Hydrocarbon Reservoir Simulation

A workflow from collecting geological data in the field to simulating and evaluating of different flow scenarios

Natural Sciences Series

Impressum / Imprint

Bibliografische Information der Deutschen Nationalbibliothek: Die Deutsche Nationalbibliothek verzeichnet diese Publikation in der Deutschen Nationalbibliografie; detaillierte bibliografische Daten sind im Internet über http://dnb.d-nb.de abrufbar.

Alle in diesem Buch genannten Marken und Produktnamen unterliegen warenzeichen-, marken- oder patentrechtlichem Schutz bzw. sind Warenzeichen oder eingetragene Warenzeichen der jeweiligen Inhaber. Die Wiedergabe von Marken, Produktnamen, Gebrauchsnamen, Handelsnamen, Warenbezeichnungen u.s.w. in diesem Werk berechtigt auch ohne besondere Kennzeichnung nicht zu der Annahme, dass solche Namen im Sinne der Warenzeichen- und Markenschutzgesetzgebung als frei zu betrachten wären und daher von jedermann benutzt werden dürften.

Bibliographic information published by the Deutsche Nationalbibliothek: The Deutsche Nationalbibliothek lists this publication in the Deutsche Nationalbibliografie; detailed bibliographic data are available in the Internet at http://dnb.d-nb.de.

Any brand names and product names mentioned in this book are subject to trademark, brand or patent protection and are trademarks or registered trademarks of their respective holders. The use of brand names, product names, common names, trade names, product descriptions etc. even without a particular marking in this work is in no way to be construed to mean that such names may be regarded as unrestricted in respect of trademark and brand protection legislation and could thus be used by anyone.

Coverbild / Cover image: www.ingimage.com

Verlag / Publisher:
AV Akademikerverlag
ist ein Imprint der / is a trademark of
OmniScriptum GmbH & Co. KG
Bahnhofstraße 28, 66111 Saarbrücken, Deutschland / Germany
Email: info@omniscriptum.com

Herstellung: siehe letzte Seite /
Printed at: see last page
ISBN: 978-3-639-49189-0

Executive Summary

This thesis is about reservoir characterization and simulation of a potential reservoir in the Ainsa Basin, Central Pyrenees, Spain.
The aim was to apply the entire workflow, from collecting geological data in the field to interpreting flow scenarios from the assumed reservoir.

In the field, column profiles of the outcrop were taken and the potential reservoir layers were defined and investigated regarding the rock properties.

A 2D CAD model was drawn with the data observed in the field. Rock and fluid property data were calculated and entered in a reservoir simulator. Also producer and injector wells were set to produce the reservoir in the most effective way.

Different flow scenarios were applied and compared, and the data collected formed the basis for the calculation of the recovery factor, the ultimate recovery and the possible reserves.

The best economic result was achieved with scenario E which has high porosity and permeability layers and represents edge-driven water drive. The corresponding pore volume is 585 m³, the oil initially in place 549 m³ and the recovery factor is 64%.
In this scenario the pressure difference is sufficient to displace the oil without gravity underride and viscous fingering.

Acknowledgements

First of all I would like thank Prof. Stephan Matthäi from the Chair of Reservoir Engineering who gave me the opportunity to work on this book. Now I have a far deeper understanding and feeling of reservoir engineering.

For the geological aspect I would like to thank Prof. R. F. Sachsenhofer and his team, namely Dr. D. Reischenbacher, J. Gusterhuber, P. Quast, S. Schnitzer and L. Scheucher, who always had time for discussion.

I also would like to express great thanks to Prof. J. Schön from the Chair of Applied Geophysics who always found time for my colleague and me to discuss our questions.

Great thanks to P. Lang who accompanied us through all the work at the computer.

Last but not least I would like to thank my colleague Albert Gasser for the intensive cooperation.

Table of content

1   Introduction 9

2   Geography 10

3   Geology 11

4   Methodology 13

   4.1   Pressure, temperature, depth 13

   4.2   Rock properties... 14

      4.2.1   General information about rock properties 14

      4.2.2   Porosity 15

      4.2.3   Permeability... 15

      4.2.4   Irreducible water saturation 15

      4.2.5   Residual oil saturation 16

      4.2.6   Capillary entry pressure 16

      4.2.7   Brooks Corey parameter 16

   4.3   Fluid properties 17

   4.4   Well placement 18

5   Results... 19

   5.1   Static results 19

      5.1.1   Calculated PV, OIIP and STOIIP... 19

      5.1.2   Monte Carlo Simulation 19

   5.2   Sweep simulation.. 21

   5.3   Production simulation... 25

   5.4   Dynamic results 32

6   Discussion 35

7   List of references 36

8   Appendix.. 37

List of figures

1: Location of the Ainsa Basin.. 10

2: Simplified and geological map Ainsa Basin... 10

3: Studied outcrop. 12

4: A, B, C Outcrop detail pictures... 12

5: CAD model with well placement.. 18

6: Expectation curve uncemented... 20

7: Expectation curve cemented... 20

8: Scenario A, sweep after 3.5 days of injection. 21

9: Scenario A, sweep after 67 days of injection. 22

10: Scenario B, sweep after 444 days of injection. 22

11: Scenario B, sweep after 1305 days of injection. 23

12: Scenario C, sweep after 10 hours of injection..................................... 23

13: Scenario C, sweep after 8 days of injection. 24

14: Scenario D, sweep after 38 days of injection. 24

15: Scenario D, sweep after 78 days of injection. 25

16: Scenario E, production simulation after 1.5 hours............................... 26

17: Scenario E, production simulation after 3 days.. 26

18: Scenario E, production simulation after 4 days. 27

19: Scenario F, production simulation after 31 hours 27

20: Scenario F, production simulation after 25 days 28

21: Scenario F, production simulation after 55 days. 28

22: Scenario G, production simulation after 1 hours 29

23: Scenario G, production simulation after 1.5 days............................... 29

24: Scenario G, production simulation after 2,5 days............................... 30

25: Scenario H, production simulation after 20 hours. 30

26: Scenario H, production simulation after 14 days. 31

27: Scenario H, production simulation after 30 days 31

28: Comparing RF, uncemented case, 60 days.. 33

29: Comparing RF, uncemented case, 10 days.. 33

30: Comparing RF, cemented case, 700 days 34

31: Comparing RF, cemented case, 100 days.. 34

32: Chart Beard and Weyl, 1973.. 37

33: Reduction of pore size due to cementation. 38

34: Capillary pressure curve (Brooks Corey Parameter). 41

35: CAD model with grid cells used for the simulator. 45

List of tables

1: Defined reservoir properties 13

2: Rock data uncemented. 14

3: Rock data cemented. 14

4: Defined fluid properties 17

5: Calculated fluid parameters. 17

6: Calculated PV, OIIP and STOIIP. 19

7: Comparison of different production scenarios. 32

8: Calculating capillary pressure curve. 41

Abbreviations

CAD	Computer Aided Design
OIIP	Oil Initially In Place
OIP	Oil In Place
STOIIP	Stock Tank Oil Initially In Place
STOIP	Stock Tank Oil In Place
RF	Recovery Factor
UR	Ultimate Recovery
PV	Pore Volume
$S_{w,irr}$	Irreducible Water Saturation
S_{or}	Residual Oil Saturation
B_o	Formation Volume Factor

Ramsauer Medardus

1 Introduction

Within a petroleum geology and reservoir engineering field study in the foreland of the Pyrenees, Spain, a large outcrop was used as a reservoir analog. This outcrop was investigated concerning petrophysical and reservoir properties, profiled, mapped and characterized in a group of two.

Column profiles and a geological cross section of the outcrop were drawn. This cross section was transformed to a CAD model, wells were set and finally the 2-D model was meshed with fine grid cells.

A simulation was performed with a 2-D reservoir simulator of the CAD model. Fluid and rock properties e.g. viscosity, density, porosity, permeability, irreducible water saturation, capillary entry pressure and other parameters were entered.
These parameters were calculated with empirical equations and physical models. Solution gas drive, gas cap drive, rock and fluid expansion drive and aquifer drive were ignored.

Different sweep and production scenarios were performed, varying in pressure gradients or cemented and uncemented reservoir layers.

During simulation, the flow behaviors like gravity underride, viscous fingering and displacement of the oil with the injection fluid could be observed.

Based on geological interpretation, rock and fluid properties, pore volume (PV), oil initially in place (OIIP) and stock tank oil initially in place (STOIIP) were calculated. Furthermore a Monte Carlo Simulation was applied to estimate proven, probable and possible reserves.

2 Geography

The investigated outcrop is located in Spain, Central Pyrenees, Ainsa Basin, Olson Member of the Escanilla Formation in the southern Buil Syncline situated between the villages Escanilla and Barcabo (Figures 1 and 2).

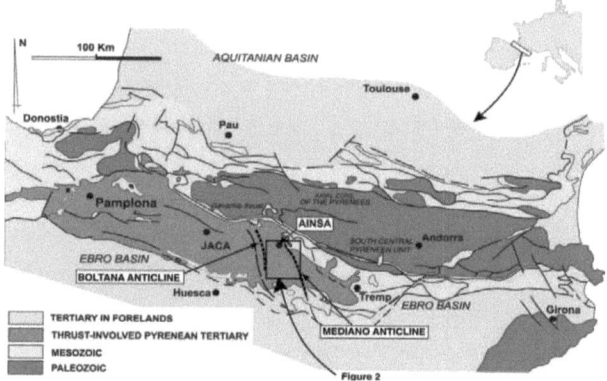

Figure 1: Location of the Ainsa Basin, Central Pyrenees, Spain (modified from Labourdette, 2011).

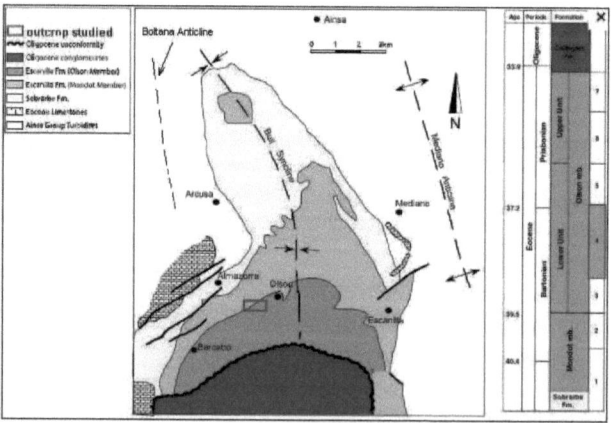

Figure 2: Simplified and modified geological map with stratigraphic column of south Ainsa Basin (Buil Syncline, modified from Labourdette, 2011). The studied outcrop is situated between Olson and Barcabo (red box), bounded to the east by the Mediano anticline and to the west by the Boltana anticline.

3 Geology

The Olson Member of the Escanilla Formation is a braided river deposit system, alternating with lateral far extended fine-grained flood plain deposits and coarse grained channel fill sequences (Figures 3 and 4).

The channel fill sequences are between 5 and 10 m in thickness. The bottom of the sequence is often coarse sandstone to conglomerate with strong contact erosion of the underlying fine-grained flood plain (Figures 3 and 4A). The grain size of the conglomerate is in the range of 2 to 15 cm, sub- to well-rounded and accumulated as cross stratification alternating with coarse sandstone layers (Figure 4B and 4C). The basis of the sequence is overlaid by coarse- to medium-grained sandstones and ends with silty shale to the top. In this silty shale bioturbation can be observed.

The flood plain deposits consist of mudstone, bioturbated siltstone and fine-grained sandstone (Figure 4D). The thickness of the layers range from decimeter to meter scale and the lateral extension is up to kilometer scale (Figure 3; Labourdette, 2011).

The studied outcrop is 187m in lateral extend, 68m in height and starts on the east side at Profile 1 and ends on the west side at Profile 3 (Figure 3). All potential reservoir rocks are conglomerates where the matrix is coarse sandstone and middle sandstone. The layers in between are silt, clay, marl and shale which are potential seal rocks. All layers have a constant dip of 12° in direction 160° (strike) and strongly react to HCL.
Due to vegetation and difficult access, no profiles have been taken on the east side of Profile 1.

The nomination of the potential reservoir layers starts with the type of the reservoir rock and the number of reservoir rock starting at the top (Figure 5 on page 19). For example "Coarse Sandstone 8" is coarse sandstone and the eighth reservoir layer from the top of the formation.
Sealing layers like silt, marl and clay are not taken into account in the nomination of the potential reservoir layers.

Ramsauer Medardus

Figure 3: Studied outcrop in the Olson Member Escanilla Formation, Ainsa Basin Spain. Three column profiles have been taken. The distance between Profile 1 and Profile 3 is 187 m and the vertical height of the reservoir at Profile 2 is 68 m.

Figure 4: A: Contact erosion of basal channel filling on the reddened flood plain. B: Basal gravel units are overlaid by coarse- to medium-grained sandstones. C: Cross stratification. D: Alternation of silty flood plains and coarse grained sandstone channel fillings.

4 Methodology

4.1 Pressure, temperature, depth

The reservoir parameters are defined in Table 1.

Reservoir parameters		
Reservoir reference depth (top of reservoir)	1700	[m]
Reservoir fluid pressure at reference depth	200	[bar]
Rock-fracture gradient	16	[kPa/m]
Geothermal gradient	25	[°C/1000m]
Water oil contact (WOC) and transition zone below the reservoir		

Table 1: Defined reservoir properties

Thus the fracture pressure at the top of the reservoir in a depth of 1700 m is 272 bar and the temperature is 57.5°C.
At the bottom of the reservoir in a depth of 1768 m, the fracture pressure is 282 bar and the temperature is 59.1°C.

4.2 Rock properties

4.2.1 General information about rock properties

The observed outcrop was strongly influenced by erosion; therefore no representative rock samples have been taken.

Rock properties were calculated with empirical correlations for sandstone (see References and Appendix). In the cemented case, the calculated values for the uncemented case were taken and porosity reduced by 50% due to cementation (see Appendix Figure 33).

Uncement layers						
Layer	Φ	k [D]	$S_{w,irr}$	r- throat [mm]	Pc [Pa]	λ
Conglomerate 1	0.20	19.6	0.018	0.21	724	3
Medium sandstone 2	0.22	0.9	0.109	0.08	1811	4
Conglomerate 3	0.16	0.2	0.119	0.08	1811	2
Medium sandstone 4	0.18	0.3	0.117	0.08	1811	3
Coarse sandstone 5	0.19	19.3	0.018	0.21	724	3
Conglomerate 6	0.19	19.2	0.018	0.21	724	3
Medium sandstone 7	0.22	0.9	0.110	0.08	1811	4
Coarse sandstone 8	0.22	33.8	0.018	0.21	724	4
Medium sandstone 9	0.22	0.9	0.110	0.08	1811	4
Coarse sandstone 10	0.22	32.8	0.018	0.21	724	4
Medium sandstone 11	0.19	0.5	0.114	0.08	1811	3
Silt	0.18	0.0	185	0.00	72426	3
Marl	0.15	0.0	19368	0.00	724264	3

Table 2: Rock data uncemented. The grey underlaid layers (2-8) are simulated. The porosity is described by Φ, the permeability by k in Darcy, $S_{w,irr}$ is the irreducible water saturation, P_e the entry pressure of the fluid in the pore space, λ the Brooks Cory parameter.

Cemented layers (50% of pore volume reduction)						
Layer	Φ	K [D]	$S_{w,irr}$	r- throat [mm]	Pc [Pa]	λ
Conglomerate 1	0.10	0.57	0.022	0.15	955	3
Medium sandstone 2	0.11	0.03	0.135	0.06	2387	4
Conglomerate 3	0.08	0.01	0.147	0.06	2387	2
Medium sandstone 4	0.09	0.01	0.144	0.06	2387	3
Coarse sandstone 5	0.10	0.56	0.022	0.15	955	3
Conglomerate 6	0.10	0.56	0.022	0.15	955	3
Medium sandstone 7	0.11	0.03	0.135	0.06	2387	4
Coarse sandstone 8	0.11	0.99	0.022	0.15	955	4
Medium sandstone 9	0.11	0.03	0.135	0.06	2387	4
Coarse sandstone 10	0.11	0.96	0.022	0.15	955	4
Medium sandstone 11	0.10	0.01	0.140	0.06	2387	3
Silt	0.09	0.00	227	0.00	95476	3
Marl	0.08	0.00	23845	0.00	954765	3

Table 3: Rock data cemented. Discription is given in Table 2.

4.2.2 Porosity

The porosity of sandstone depends strongly on sorting and grain size which was estimated in the field with the help of millimeter scale and comparison charts.

With these two parameters the uncompacted surface porosity was defined by the chart of Beard and Weyl (1973). See Figure 32 in the Appendix.

The compacted porosity is calculated with the empirical correlation of Athy (1930). For the uncemented case, calculated values of the porosity are taken for the simulation (Tables 2 and 3).

A 50 % pore space reduction equates to a cement thickness of 10% of the grain diameter. The result of the uncemented and cemented case is a porosity varying between 8 and 22 %.

For charts, equations and figures, see Appendix, Chapter 8.1

4.2.3 Permeability

The permeability of the sandstone layers is calculated with the correlation of Berg (1970). For this equation, porosity, grain diameter and the sorting parameters are required.

The results show that permeability is strongly dependent on porosity but also on grain diameter and sorting. The permeability is in the range from 30 md to 33.8 D (Tables 2 and 3)

For the Berg equation see Appendix, Chapter 8.2

4.2.4 Irreducible water saturation

To calculate the irreducible water saturation Timur's equation (1968) is transformed to S_{wirr} (see Appendix, Chapter 8.3).

Depending on permeability and porosity, S_{wirr} ranges from 1.8 and 14.7% (Tables 2 and 3).

4.2.5 Residual oil saturation

The definition of the residual oil saturation (S_{or}) is challenging because of lack of proper information.

Based on measurements and statistical distributions of the frio fluvial-deltaic sandstone pay in South Texas, the S_{or} of 30% was adapted for the Olson Member (McRae et al., 1995).

4.2.6 Capillary entry pressure

The capillary pressure is calculated with Equation (4), Appendix 8.4. The throat radius can easily be calculated with the equation of Pythagoras. The entry pressure depends on the interfacial tension (IFT) and the wetting angle. Both of these are dependent on the lithology and fluid parameters.

The capillary entry pressure is between 724 Pa for coarse sandstone uncemented and 954765 Pa for marl (Tables 2 and 3).

For equation and parameters, see Appendix, Chapter 8.4

4.2.7 Brooks-Corey parameter

The Brooks-Corey parameter (BC) results from the sorting of a siliciclastic sandstone. Bad sorting (inhomogeneous grain size distribution) leads to high transition zones, good sorting to small transition zones.

The values for BC are 5 for very good sorting and 1 for very bad sorting (Li, 2004).

For more detail, see Appendix, Chapter 8.5.

4.3 Fluid properties

The fluid parameters are defined in Table 4.

Fluid parameters			
API grad at standard conditions	API_{SC}	35	[°]
Oil bubble point at reservoir conditions	p_{bRC}	30	[bar]
Oil viscosity	μ_{SC}	7	[cP]
Salinity injection water		20000	[ppm]

Table 4: Defined fluid properties

Furthermore fluid parameters like formation volume factor (B_o), viscosity (μ) of oil and water and density (ρ) of oil and water under reservoir conditions have been calculated (Table 5).

For equations used in these calculations, see Appendix, Chapter 8.6.

Calculated fluid parameter		
B_o	1	[-]
$\mu_{od,res}$	3.64	[cP]
$\mu_{w,res}$	0.48	[cP]
$\rho_{o,res}$	838	[kg/m³]
$\rho_{w,sal,res}$	1006.31	[kg/m³]

Table 5: Calculated fluid parameters. Bo is the formation volume factor, $\mu_{od,res}$ the viscosity of the dead oil under reservoir conditions (RC), $\mu_{w,res}$ the viscosity of the water under RC, $\rho_{od,res}$ density of the dead oil under RC and $\rho_{w,res}$ density of the water under RC.

4.4 Well placement

As displayed in Figure 5, only the reservoir layers Coarse Sandstone 5, Conglomerate 6, Medium Sandstone 7 und Coarse Sandstone 8 can be found throughout the entire reservoir. Hence the production well (Well 1) is placed in the middle of the formation and deviated to get access to all layers.

On the left side (east, Well 2) and on the right side (west, Well 3) injection wells are placed.

Each reservoir layer is perforated at all three wells.

The perforation is named by the number of the well and the name of the layer. For example "Well 38" means the perforation of Well 3 (Injector) at layer 8.

The grid cells in Figure 5 are implemented. The simulation of the reservoir was performed with 15337 cells and 7808 points (see Appendix 6.8).

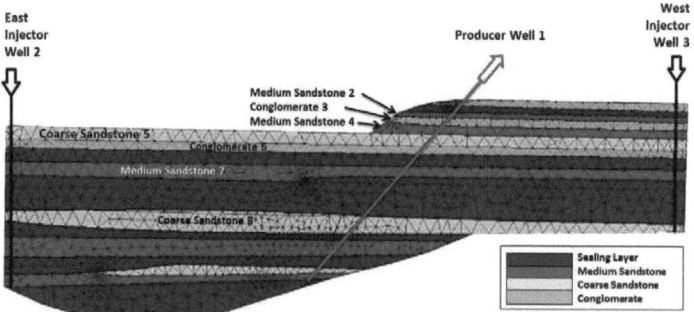

Figure 5: CAD model with well placement and implemented grid cells. The lateral extension of the reservoir is 187 m and the height is 68 m.

5 Results

5.1 Static results

5.1.1 Calculated pore volume (PV), oil initially in place (OIIP) and stock tank oil initially in place (STOIIP)

Table 6 displays the calculated values of pore volume, oil initialyl in place and stock tank oil initially in place for the uncemented reservoir and for the cemented reservoir (50 % pore space reduction due to cementation of the grains with a thickness of 10 % of the grain diameter). The values are the sum of all simulated oil bearing layers from medium-grained sandstone 2 to coarse grained sandstone 8.

For equations, see Appendix 8.7.

	Res. Volume [m³]	PV [m³]	OIIP [m³]	STOIIP [m³]
Uncemented Reservoir	2866	585	549	548
Cemented Reservoir	2866	292	276	276

Table 6: Calculated values for pore volume (PV), oil initially in place (OIIP) and stock tank oil initially in place (STOIIP).

5.1.2 Monte Carlo Simulation

To estimate the probability of the reserves, a Monte Carlo (MC) Simulation was run.

Normal distributions for porosity (Φ) and the volume of the oil bearing layers (V) were generated.

The normal distribution of the porosity was truncated at a range of ± 10 % and the volume at a range of ± 15 %.

The irreducible water saturation depends on the porosity (calculated with transformed Timur equation).

The results of the Monte Carlo Simulation display proven (with 90 % probability), probable (50 %) and possible (10 %) reserves (Figures 26 and 27).

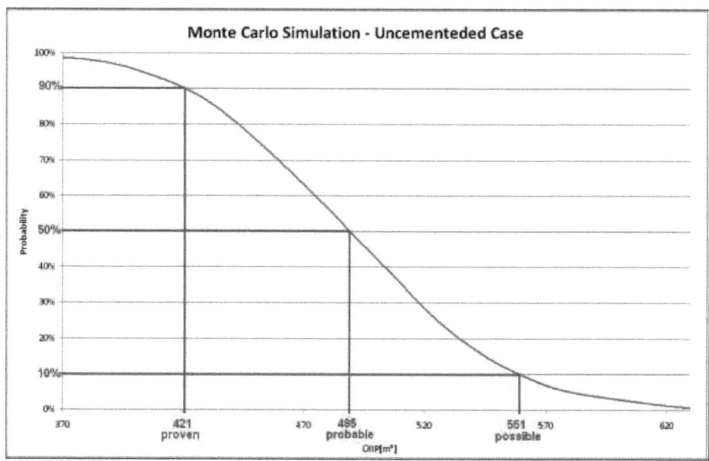

Figure 6: Expectation curve uncemented to estimate the probability of oil initially in place (OIIP).

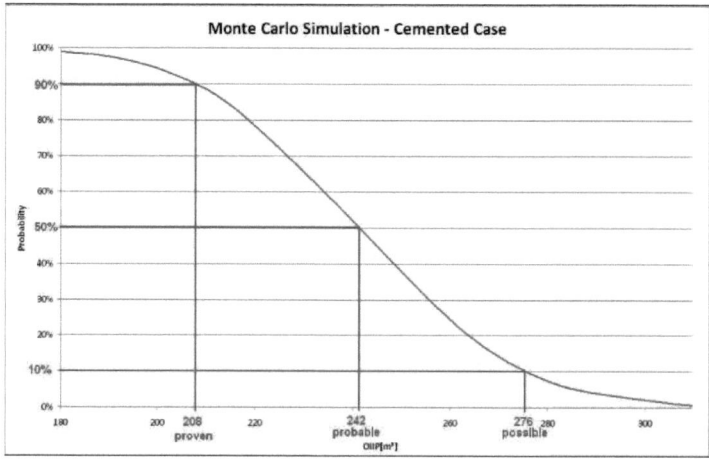

Figure 7: Expectation curve cemented to estimate the probability of oil initially in place (OIIP).

Compared with the static calculated values (Table 6), the Monte Carlo Simulation shows slightly lower values for oil initially in place (OIIP). The probable values (with 50 % probability) are about 12 % lower than the calculated values from Table 6. The possible values, which may occur with a probability of 10 %, correlate with the calculated values of Table 6.

5.2 Sweep simulation

Four different sweep simulations were performed and are being discussed in this chapter.

Scenario A: Pressure difference 9 bar, uncemented.
Scenario B: Pressure difference 9 bar, cemented.
Scenario C: Pressure difference 90 bar, uncemented.
Scenario D: Pressure difference 90 bar, cemented.

In case of a sweep simulation the heterogeneity in terms of fluid flow of the reservoir can be observed because all layers are charged with the same pressure. Sweep direction is from left to right.

Scenario A: Pressure difference 9 bar, uncemented.

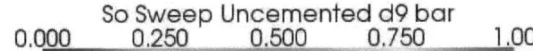

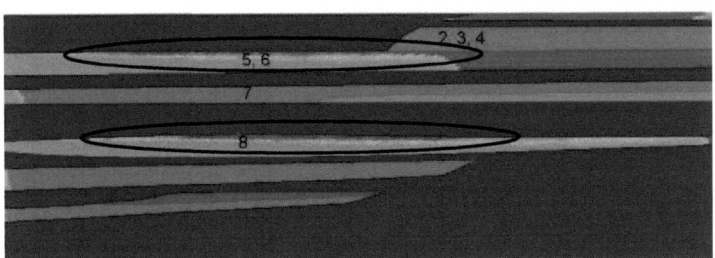

Figure 8: Scenario A, sweep uncemented, 9 bar p-difference. Oil saturation (S_o) after 3.5 days of injection.

In this scenario, the first water breakthrough occurs after 3.5 days in layer 8 (Figure). Strong gravity underride in layers 5, 6 and 8 (black circle) can be identified.

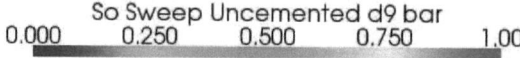

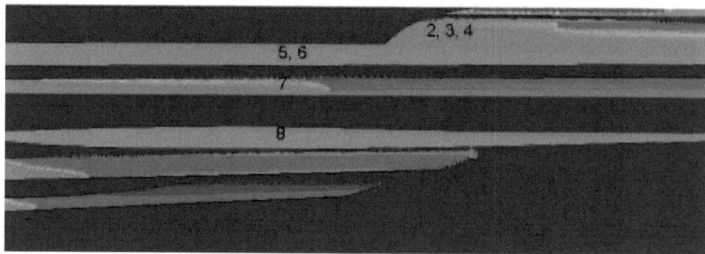

Figure 9: Scenario A, sweep, uncemented, 9 bar p-difference. Oil saturation (S$_o$) after 67 days of injection.

After 67 days of injection in layers 5,6 and 8, the gravity underride shows no effect anymore (Figure 9).

Scenario B: Pressure difference 9 bar, cemented.

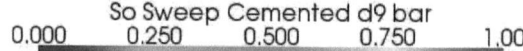

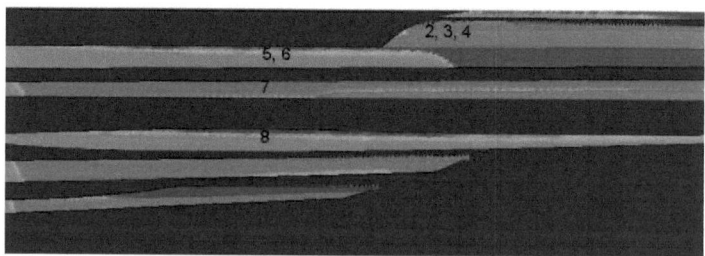

Figure 10: Scenario B, sweep, cemented, 9 bar p-difference. Oil saturation (S$_o$) after 444 days of injection.

In the cemented scenario, the first water breakthrough in layer 8 occurs after 444 days. Strong gravity underride can be seen as in Scenario A (uncemented), but with a significant delay (Figure 10).

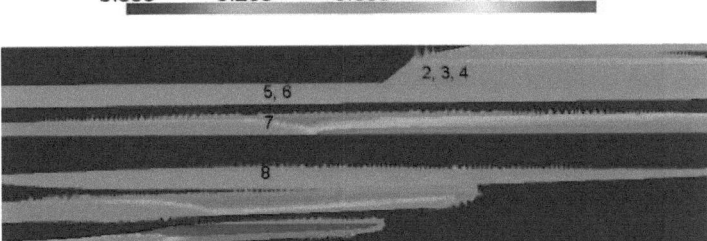

Figure 11: Scenario B, sweep, cemented, 9 bar p-difference. Oil saturation (S_o) after 1305 days of injection.

Scenario C: Pressure difference 90 bar, uncemented.

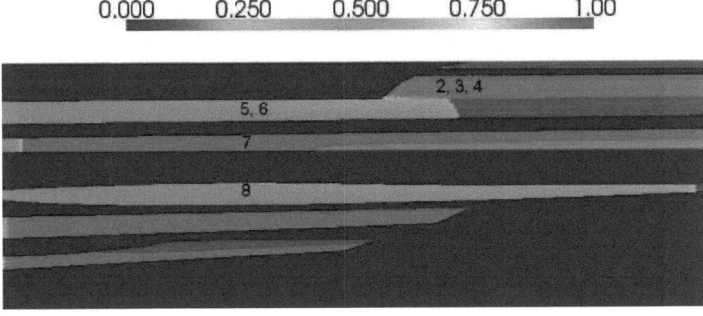

Figure 12: Scenario C, sweep, uncemented, 90 bar p-difference. Oil saturation (S_o) after 10 hours of injection.

In comparison to Scenario A, the pressure difference is ten times higher. As a result no gravity underride can be seen (Figure 12 and 13).

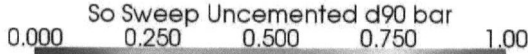

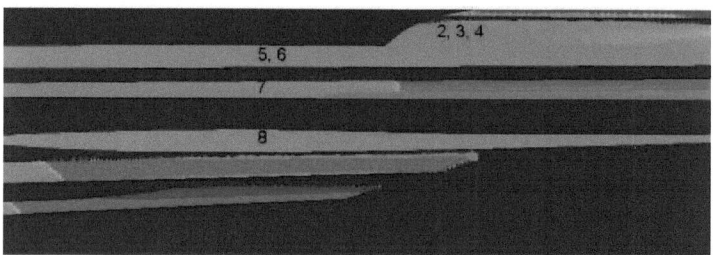

Figure 13: Scenario C, sweep, uncemented, 90 bar p-difference. Oil saturation (S_o) after 8 days of injection.

Scenario D: Pressure difference 90 bar, cemented.

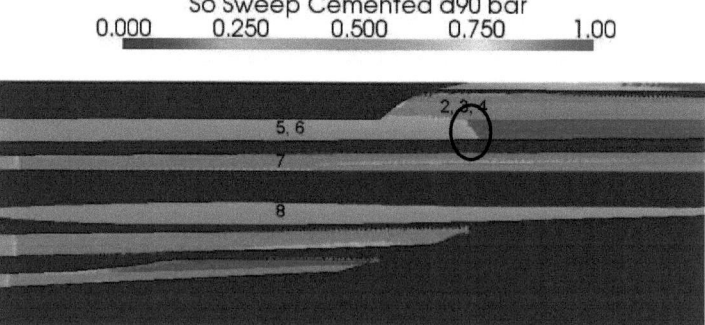

Figure 14: Scenario D, sweep, cemented, 90 bar p-difference. Oil saturation (S_o) after 38 days of injection.

As it can be seen in Figure 14 (black circle), gravity underride can be observed in the higher permeable layers, but no viscous fingering.

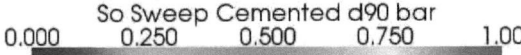

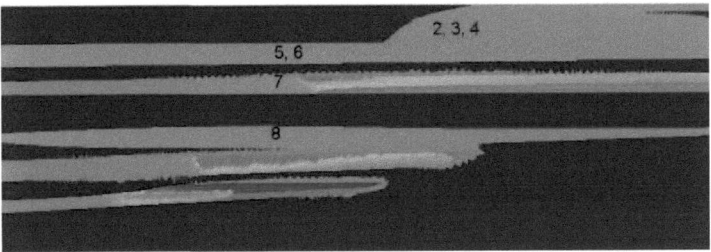

So Sweep Cemented d90 bar

0.000   0.250   0.500   0.750   1.00

Figure 15: Scenario D, sweep, cemented, 90 bar p-difference. Oil saturation (S$_o$) after 78 days of injection.

5.3 Production simulation

Four different production simulations have been performed.

Scenario E: Injection pressure 200bar, uncemented.

Scenario F: Injection pressure 200bar, cemented.

Scenario G: Injection pressure 265bar, uncemented.

Scenario H: Injection pressure 265bar, cemented.

200 bar relates to the given reservoir pressure. Due to the fact that the simulator cannot simulate water drive (edge driven from the left and right side), the production with injection wells was simulated.

265 bar is the maximum allowable injection pressure calculated from the fracture pressure of the rock with a 3 % safety margin.

For all cases, the well flowing pressure is 100 bar, the production well is in the middle of the reservoir (red lines) and the injection wells are on the left (east) and the right (west) side (black lines, Figure 5).

The left (east) well is injecting into layer 5 to 8 and the right (west) well into layer 1 to 8. The middle well is producing from layers 1 to 8.

All oil bearing layers are fully saturated with oil, only the irreducible water is in the pore space.

Seal layers (clay, marl, silt) are saturated with water.

For oil initially in place (OIIP), just the simulated layers are taken into account.

All four simulations were stopped after ten days of simulator runtime.

Fluid parameters can be seen in Tables 4 and 5. For the rock properties, see Tables 2 and 3.

Scenario E: Injection pressure 200bar, uncemented.

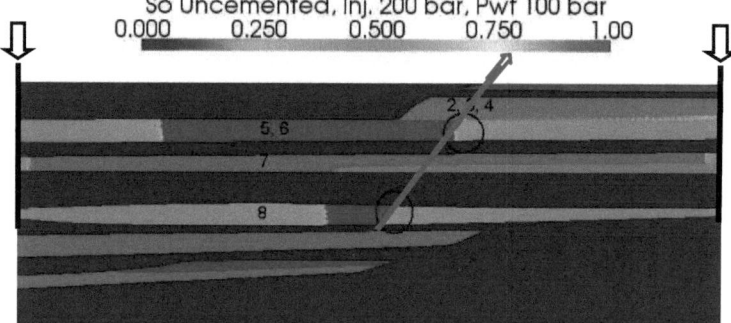

Figure 16: Scenario E, production simulation, uncemented, Inj-p 200 bar, oil saturation (S_o). First water breakthrough in layers 5, 6 and 8 after 1.5 hours; RF= 17%.

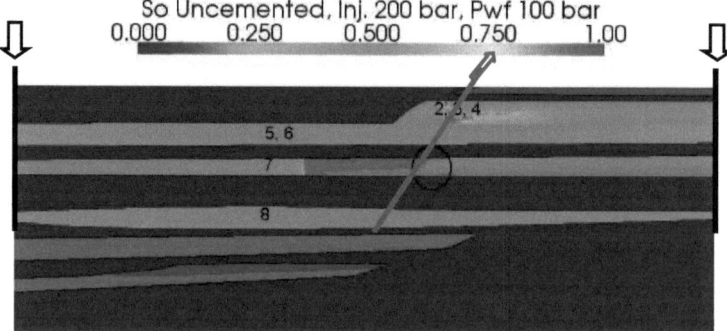

Figure 17: Scenario E, production simulation, uncemented, Inj-p 200 bar, oil saturation (S_o). Water breakthrough in layer 7 after 3 days. No gravity underride can be observed; RF=53%.

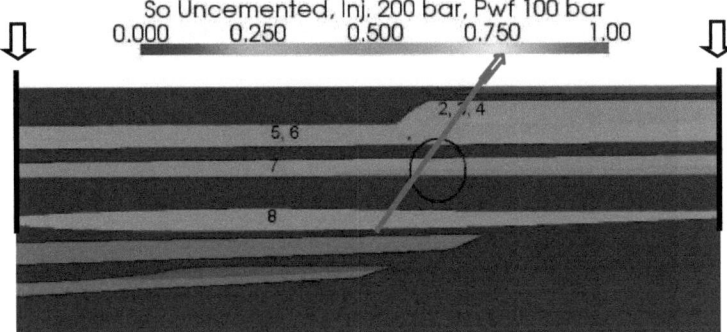

Figure 18: Scenario E, production simulation, uncemented, Inj-p 200 bar, oil saturation (S$_o$). The last water breakthrough occurs in layer 7 after 4 days of injection; RF=56%.

The simulation of Scenario E was stopped after 176 days of simulation with a recovery factor of 65%.

Scenario F: Injection pressure 200bar, cemented.

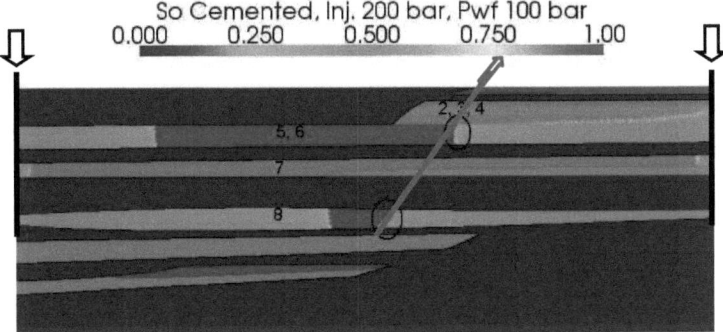

Figure 19: Scenario F, production simulation, cemented,Inj-p 200 bar, oil saturation (S$_o$). First water breakthrough in layers 5,6 and 8 after 31 hours; RF= 20%.

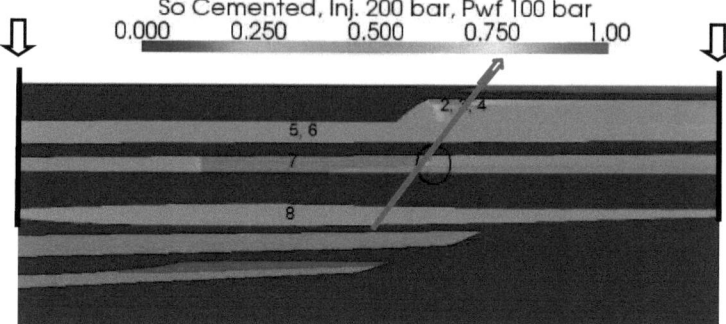

Figure 20: Scenario F, production simulation, cemented, Inj-p 200 bar, oil saturation (S_o). Water breakthrough in layer 7 after 25 days (circle). No gravity underride can be observed; RF=49%.

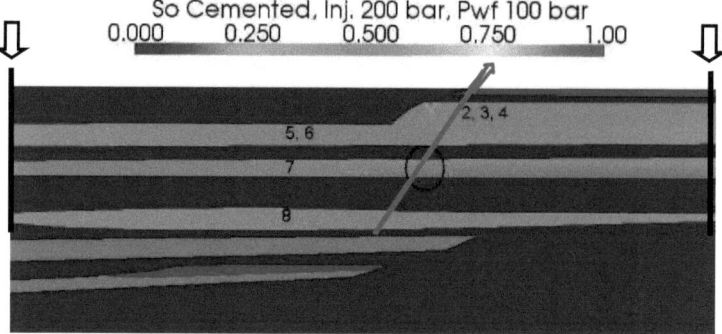

Figure 21: Scenario F, production simulation, cemented, Inj-p 200 bar, oil saturation (S_o). The last water breakthrough occurs in layer 7 after 55 days of injection (circle); RF=56%.

The simulation was stopped after 5192 days (14.2 years) of simulated time and at a recovery factor of 67%.

Compared with the uncemented case (Scenario E), a recovery factor of 65% is gained after 3.6 years (176 days in the uncemented case).

Scenario G: Injection pressure 265 bar, uncemented.

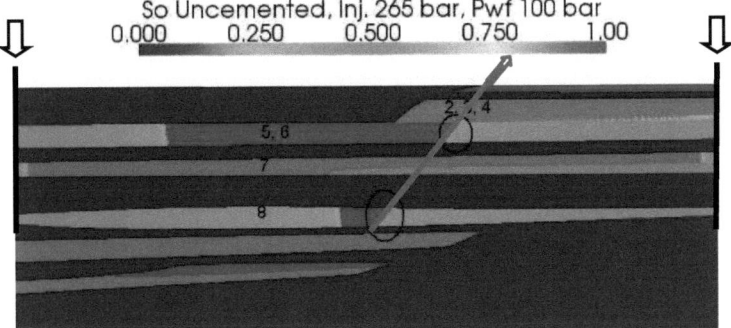

Figure 22: Scenario G, production simulation, uncemented, Inj-p 265 bar, oil saturation (S_o). First water breakthrough in layers 5.6 and 8 after 1 hours (circle); RF= 16%.

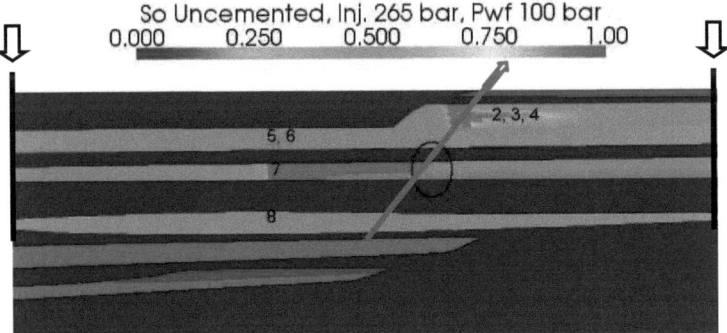

Figure 23: Scenario G, production simulation, uncemented, Inj-p 265 bar, oil saturation (S_o). Water breakthrough in layer 7 after 1.5 days (circle). No gravity underride can be observed; RF=50%.

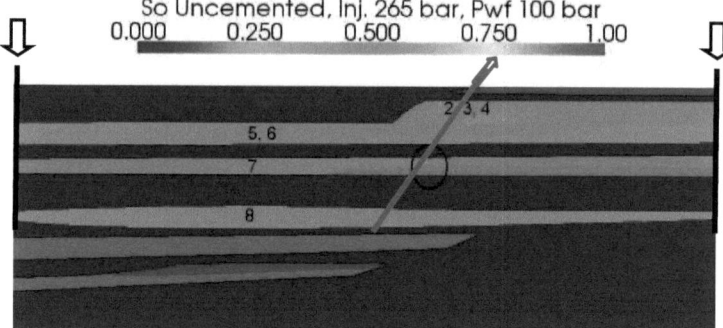

Figure 24: Scenario G, production simulation, uncemented, Inj-p 265 bar, oil saturation(S_o). The last water breakthrough occurs in layer 7 after 2.5 days of injection (circle); RF=54%.

The simulation was stopped after 60 days of simulated time with a recovery factor of 64.55%.

Scenario H: Injection pressure 265 bar, cemented.

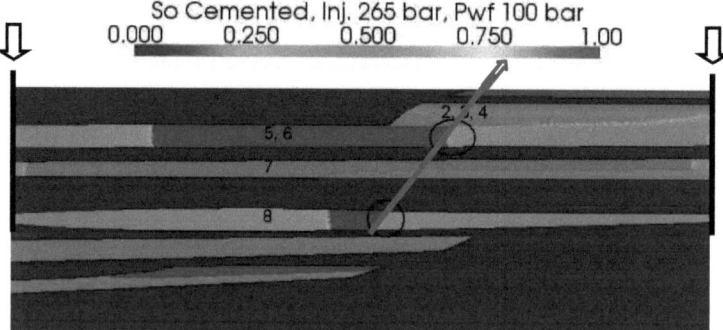

Figure 25: Scenario H, production simulation, cemented, Inj-p 265 bar, oil saturation (S_o). First water breakthrough in layers 5.6 and 8 after 20 hours (circle); RF= 21%.

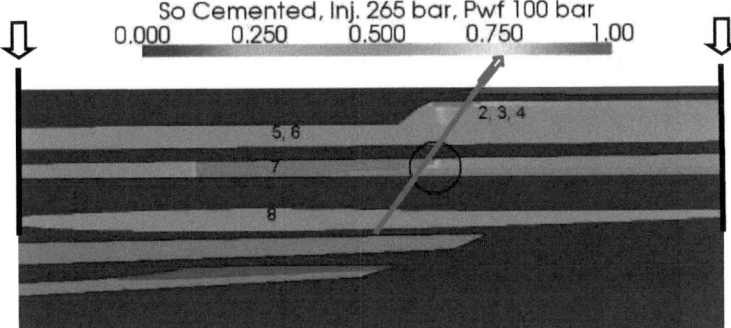

Figure 26: Scenario H, production simulation, cemented, Inj-p 265 bar, oil saturation (S_o). Water breakthrough in layer 7 after 14 days (circle). No gravity underride can be observed; RF=49%.

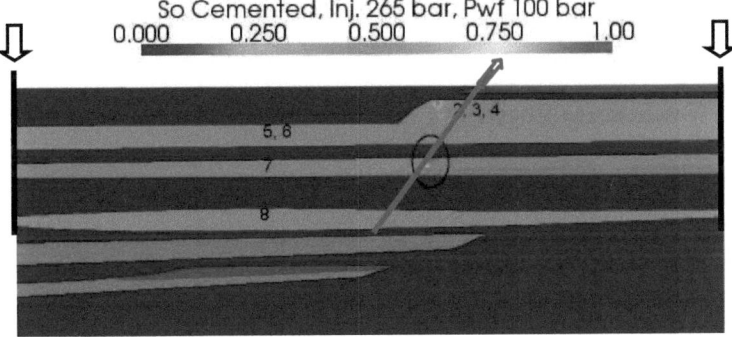

Figure 27: Scenario H, production simulation, cemented, Inj-p 265 bar, oil saturation (S_o). The last water breakthrough occurs in layer 7 after 30 days of injection (circle); RF=55%.

The simulation was stopped after 1720 days of simulated time with a recovery factor of 66.1%.

In comparison to the uncemented case (Scenario G), a recovery factor of 64 % is reached after 328 days (39 days in the uncemented case).

5.4 Dynamic results

The dynamic results are calculated with the output data given by the reservoir simulator which are a function over time.

Comparison of different scenarios		RF in dec.% after		UR [m³]
	Injection- p [bar]	58 days	694 days	
Sc. E, Uncemented	200	0.643		353
Sc. G, Uncemented	265	0.645		354
Sc. F, Cemented	200		0.645	178
Sc. H, Cemented	265		0.653	180

Table 7: Comparison of different production scenarios (Sc.) in terms of recovery factor (RF) and ultimate recovery (UR).

As it can be seen in Table 7, slightly higher recovery factors are achieved due to higher injection pressure, hence the ultimate recovery.

Because of low computer performance and the lack of time and resources, the simulation was canceled after 10 days of continuous running, resulting in 58 days of simulated time for the uncemented cases and 694 days for the cemented cases.

It has to be considered that only the simulated layers are counted to among the OIIP. All other layers like silt, marl, and clay are saturated with water.

Figures 28 to 31 display the recovery factor and the speed of depletion over time for Scenarios E-H.

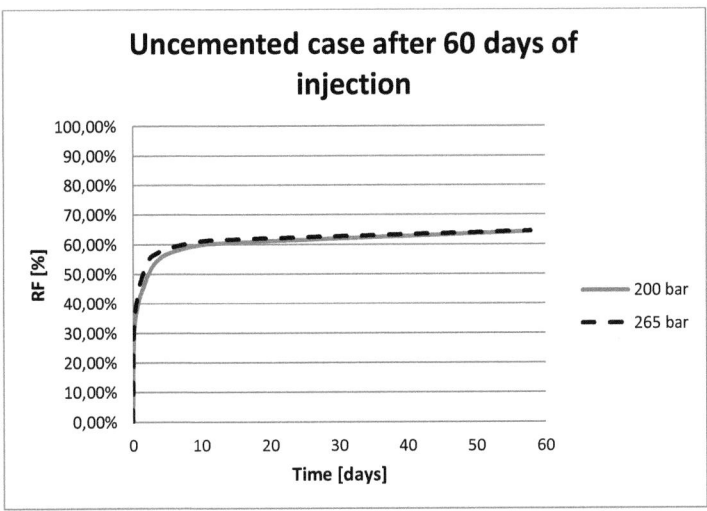

Figure 28: Comparing uncemented case with 200 bar and 265 bar of injection pressure in terms of recovery factor after 60 days of injection.

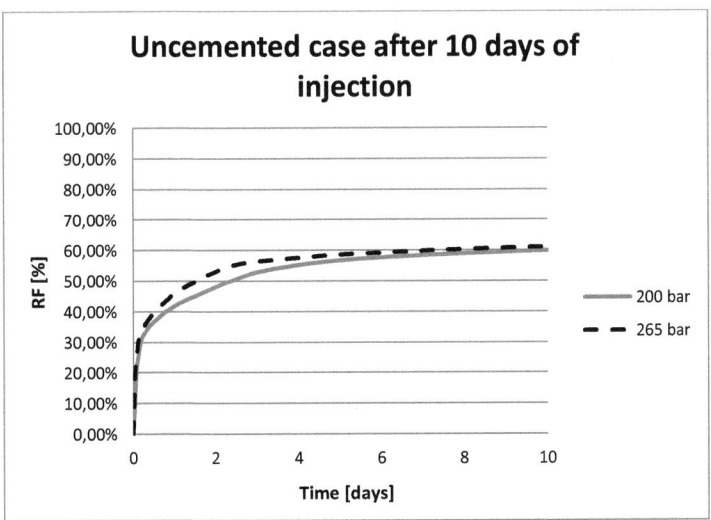

Figure 29: The same conditions as in figure 28 but displaying recovery behavior in more detail within the first 10 days of injection.

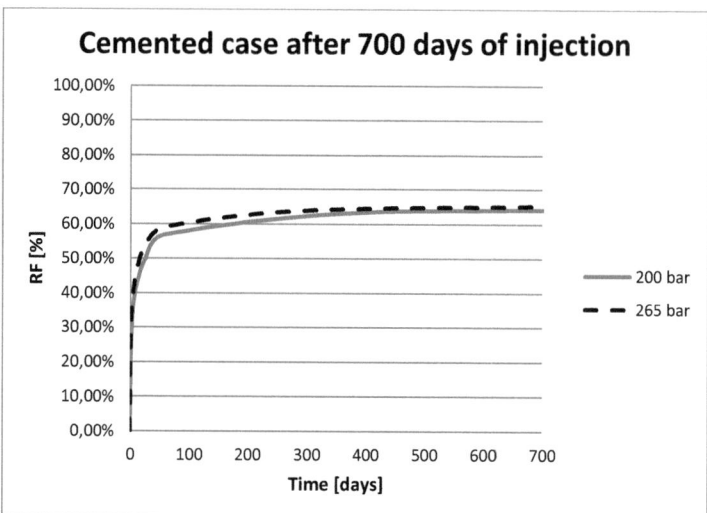

Figure 30: Comparing cemented case with 200 bar and 265 bar of injection pressure in terms of recovery factor after 700 days of injection.

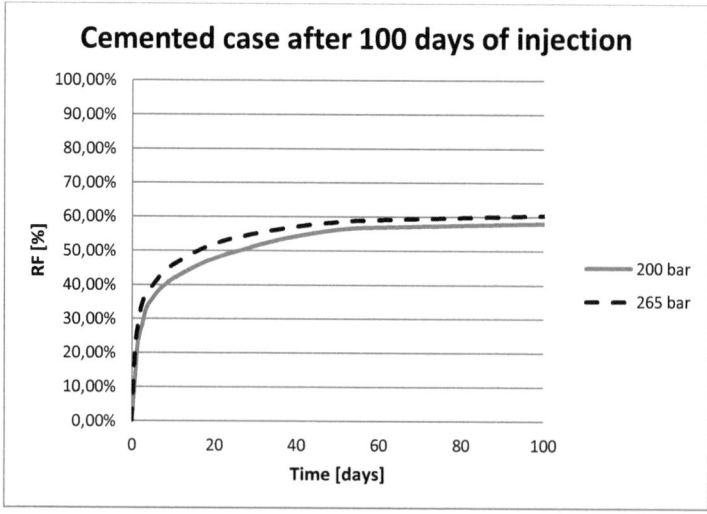

Figure 31: The same conditions as in figure 30 but displaying recovery behavior in more detail within the first 100 days of injection.

6 Discussion

The result of the sweep simulation shows that a minimum pressure gradient is essential for good displacement of the oil. Otherwise gravity underride and viscous fingering can occur resulting in trapped, hence non-producible oil (see Figure 10).

The investigation of the production scenarios reveals that higher injection pressure increases the recovery factor only to a very small degree (0.2% in the uncemented case and 0.8% in the cemented case). It is a question of costs and revenue whether injection is economic or not and has to be evaluated.

Scenarios E and F represent the water drive mechanism with the given reservoir pressure of 200 bar but the injection pressure stays constant over time.

In this type of reservoir (long lateral extension of the reservoir layers) it can easily occur that the edge driven water drive mechanism stays constant at a high pressure long enough to deplete the reservoir sufficiently. In that case the installation of injection wells cannot be recommended.

The biggest variation of OIIP and RF is a consequence of the degree of cementation and residual oil saturation.
If the cement layer thickness is only 10 % of the grain diameter, the pore space is reduced by 50 % (see Figure) and by even a layer thickness of 20 % of the grain diameter, the reduction in pores space is 80%.
Without sufficient representative core analysis, no significant simulation and prediction of OIIP can be carried out.

7 List of references

Athy L. F., 1930. Density, porosity and compactation of sedimentary rocks, *AAPG Bull.,* 14, pp 1-24.

Beard, D. C. and Weyl, P. K., 1973. Influence of Texture on Porosity and Permeability of Unconsolidated Sand, *AAPG Bull.,* 57 (2), pp 349-369.

Beggs H.D. and Robinson J.R., 1975. Estimating the Viscosity of Crude Oil Systems, *JPT* (9), pp1140-41.

Berg, R. R., 1970. Method of determining permeability from reservoir rock properties, *Trans. Gulf Coast Ass. of Geol. Soc.,* 20, pp 303-317

Hanafy H.H., Macary S.M., ElNady Y.M., Bayomi A.A. and El Batanony M.H., 1997. A New Approach for Predicting the Crude Oil Properties, *SPE 37439.*

Labourdette R., 2011. Stratigraphy and static connectivity of braided fluvial deposits of the lower Escanilla Formation, south central Pyrenees, Spain. *AAPG Bull,* 95 (4), pp. 585-617.

Li, Kewen, 2004. Theoretical Development of Brooks Corey Capillary Pressure Model from Fractal Modeling of Porous Media. *SPE 89429*

McRae L. E. el al, 1995. Revitalizing a Mature Oil Play Strategies for Finding and Producing Unrecovered Oil in Frio Fluvial- Deltic Sandstone Reservoirs of South Texas. *University of Texas at Austin. Austin, TX* pp.78713- 8924.

Timur, A., 1968. An investigation of permeability, porosity, and residual water saturation relationships for sandstone reservoirs: *The Log Analyst,* 9 (4), pp 8-17.

8 Appendix

8.1 Porosity

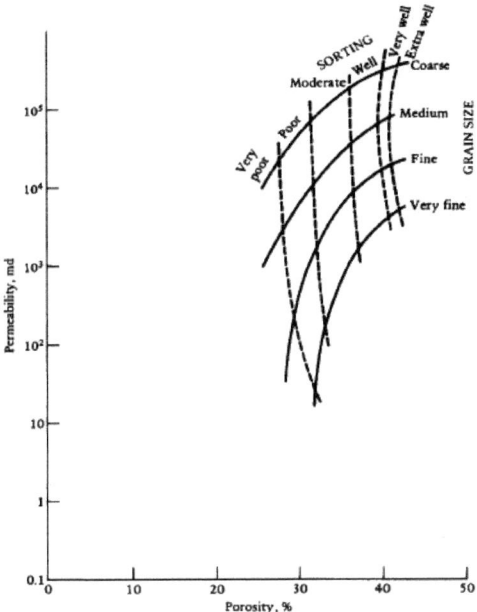

Figure 32: Estimating porosity and permeability of unconsolidated sand (Beard and Weyl, 1973). In this case just the porosity is required.

Calculating porosity of a given rock type in a certain depth with the empirical Athy equation for siliciclastic rocks (1).

$$\phi_{(z)} = \phi_0 * e^{(-b*z)} \qquad (1)$$

ϕ_0 is the surface porosity,

b is the rock skeleton factor and

z is the depth of interest.

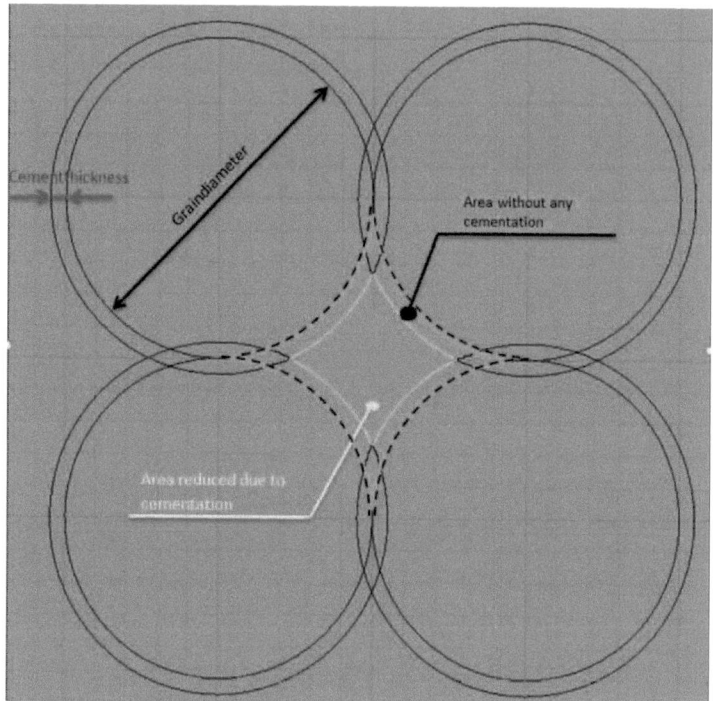

Figure 33: Reduction of pore size due to cementation.

A 10 % diameter increase due to cementation results in a 50 % pore space reduction.

The area within the dashed line is the flow area without any cementation.

The area within the yellow rimmed line is the flow area with cementation.

8.2 Permeability

For calculating the permeability, the empirical equation of Berg (1973) is used (2).

$$k = 5.1 * 10^{-6} * \phi^{5.1} * d^2 * e^{(-1.385*\psi)} \qquad (2)$$

k is the permeability in darcy,

ϕ the porosity in percent,

d the grain diameter in mm and

ψ the sorting in phi units.

As it can be seen in the equation, permeability is strongly depends on porosity.

8.3 Irreducible water saturation

In order to calculate the irreducible water saturation, the Timur equation is transformed to S_{wirr} (3).

$$k = 100 * \left[\frac{\phi^{2.25}}{S_{w,irr}}\right]^2 \qquad (3)$$

k is the permeability in md

ϕ the porosity and

S_{wirr} in dec %.

8.4 Capillary entry pressure

Equation (4) for calculating the capillary entry pressure.

$$P_c = \frac{2\sigma * \cos(\theta)}{r} \qquad (4)$$

σ is the wetting angle,
θ is the interfacial tension (IFT) and
r is the pore throat radius.

Quartz-sandstone has a wetting angle of 0° and for the IFT 75 dynes / cm are taken. The pore throat radius can easily be calculated with the equation of Pythagoras.

8.5 Brooks-Corey (BC) parameter

Li (2004) describes the behavior of the Brooks-Corey parameter and uses
equation (5).

$$P_c = p_c * (S_w^*)^{\frac{1}{\lambda}} \qquad (5)$$

p_c is the entry pressure,

λ is the Brooks-Corey parameter,

S_w^* is the normalized saturation of the wetting phase.

This leads to Brooks-Corey curves shown in Figure 32.

Entry pressure [Pa]	724
BC1 [-]	1
BC2 [-]	5

Sw [dec%]	Capillary pressure (BC1) [Pa]	Capillary pressure (BC2) [Pa]
0,1	7240	1147
0,2	3620	999
0,3	2413	921
0,4	1810	870
0,5	1448	832
0,6	1207	802
0,7	1034	778
0,8	905	757
0,9	804	739

Table 8: Calculating capillary pressure curve depending on entry pressure, actual
water saturation and Brooks-Cory parameter.

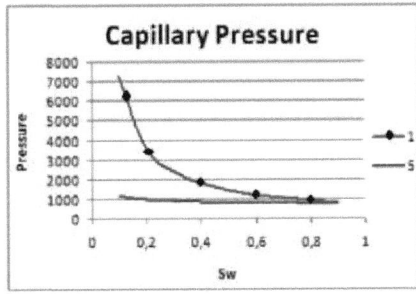

Figure 34: Capillary pressure curve (Brooks-Corey parameter).

8.6 Fluid Properties

The oil formation volume factor B_o is calculated with the Standing Equation
(6).

$$B_o = 0.972 + 0.00825 * \left\{ R_s * \left(\frac{\rho_g}{\rho_o}\right)^{0.5} + 7.124 + 0.01 * T \right\}^{1.175} \quad (6)$$

and

$$B_o = 1 + 0.008 * R_s * \left(\frac{\rho_g}{\rho_o}\right)^{0.5} + \frac{0.011}{(\rho_g * \rho_o)} * (0.063 * T - 1) \quad (7)$$

Levitan Equation (7).

B_o is the formation volume factor
R_s is the solution gas oil ratio,
ρ_g is the density of the gas in [kg/m³],
ρ_o is the density of the oil in [kg/m³],
T is the temperature in Kelvin.

Both equations deliver a formation volume factor of one.

For calculating the dead oil viscosity at reservoir conditions the arithmetic mean of the equation of Hanafy et al. (1997), (8)

$$\mu_{ob} = e^{[7.296 * \rho_{ob}^3 - 3.095]} \qquad (8)$$

μ_{ob} is the viscosity of the oil at bubble point,
ρ_{ob} is the density of the oil at bubble point.

and

The equation of Beggs and Robinson (1975), (9) were used.

$$\mu_{OD} = 10^X - 1 \qquad (9)$$

Where

$$X = YT^{-1.163}$$

$$Y = 10^Z$$

$$Z = 3.0324 - 0.02023 * API$$

μ_{OD} is the dead oil viscosity,
T the temperature in Fahrenheit and
API in API-Grad.

Equation (8) calculates the oil viscosity at bubble point and equation (9) the dead oil viscosity.

8.7 Calculation of pore volume (PV), oil initially in place (OIIP), stock tank oil initially in place (STOIP), recovery factor (RF) and ultimate recovery (UR)

Equation (10) for calculating the pore volume.

$$PV = V * \phi \qquad (10)$$

PV is the pore volume,

V is the volume of the oil bearing layers and

ϕ is the porosity.

Equation (11) calculating oil initially in place.

$$OIIP = PV * \left(1 - S_{w,irr}\right) \qquad (11)$$

OIIP is the oil initial in place,

PV is the pore volume and

$S_{w,irr}$ is the irreducible water saturation.

Equation 1: Calculating stock tank oil initial in place.

$$STOIIP = \frac{OIIP}{B_o} \qquad (12)$$

STOIIP is the stock tank oil initially in place,

OIIP is the oil initial in place and

B_o is the formation volume factor.

Equation 2: Calculating recovery factor.

$$RF = \frac{OIIP}{(1-OIP)}$$
(13)

RF is the recovery factor,

OIIP is the oil initially in place and

OIP is the oil in place.

Equation 3: Calculating ultimate recovery

$$UR = STOIIP * RF$$
(14)

UR is the ultimate recovery,

STOIIP is the stock tank oil initially in place and

RF is the recovery factor.

8.8 CAD model with grid cells

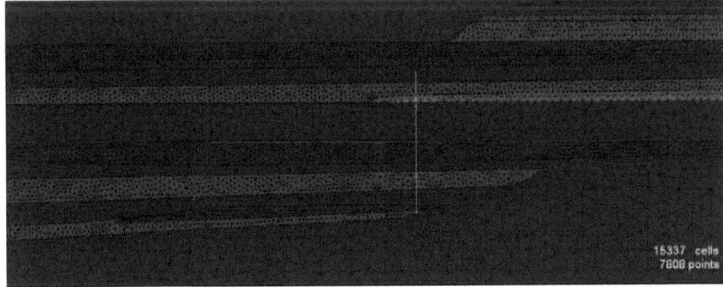

Figure 35: CAD model with grid cells used for the simulator. The lateral extension of the reservoir is 187 m and the height is 68 m.

The reservoir was split in 15337 cells and 7808 points.

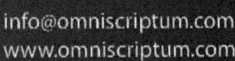

Printed by Books on Demand GmbH, Norderstedt / Germany